Ten Essential Skills for Calculus:
A Guide for the Prospective Student

by

Richard Shedenhelm

CONTENTS

INTRODUCTION

Many years ago, starting to tutor college students in calculus, I quickly realized that a large percentage of what I was instructing was not calculus at all but instead precalculus topics. The years since have only repeated the same observation. This phenomenon may be the result of many students not taking calculus immediately after their precalculus course. In either case, the purpose of this work is to identify the top ten skills that are crucial for the success of the calculus student.

The ten skills are: Factoring/Expanding, Fractions, Exponents/Radicals, Algebraic Mal-Rules, Difference Quotients, Functions, Graphs, Linear Functions, Plane Geometry, and Trigonometry. In this book, the section on trigonometry will be the longest, but the material needed by the student is much less than might appear to be the case. The length of this latter section is due in part to my conviction that trigonometric principles need the most explanation of any of the ten areas.

Athens, Georgia, March 5, 2015.

Richard Shedenhelm

1. Factoring and Expanding

The quantities which, when added (or subtracted) together, produce a quantity, are called the *terms* of the quantity. For example, x^2, $5x$, and 6 are the terms of $x^2 + 5x - 6$. The quantities which, when multiplied together, produce a quantity, are called the *factors* of the quantity. For example, 5 and x are the factors of $5x$. Keeping the conceptual distinction between "terms" and "factors" is crucial to the mathematics student since the properties of addition and multiplication have important algebraic differences.

  Factoring is the process of splitting an expression of two or more terms into an equivalent product of two or more factors. For example, $x^2 + 5x - 6$ can be factored into $(x + 6)(x - 1)$. There are five factoring principles the calculus student must know: 1. Greatest Common Factor; 2. Difference of Squares; 3. Sum/Difference of Cubes; 4. Perfect Square Trinomials; 5. Completing the Square.

  1. Greatest Common Factor. Always be on the lookout for terms that have a common factor. This factoring technique is very common and powerful. For example, the GCF of $4x^3 + 6x^2 - 10x$ is $2x$. Hence, when factored we have $2x(2x^2 + 3x - 5)$. Even if you fail to identify the *greatest* common factor, getting the expression into any sort of factored form is a step in a right direction.

  2. Difference of Squares. This is the next most important factoring principle. This technique can be summarized as

$$a^2 - b^2 = (a + b)(a - b)$$

Examples:

$$x^2 - 1 = x^2 - 1^2 = (x + 1)(x - 1)$$

$$x^6 - x^4 = x^4(x^2 - 1) = x^4(x + 1)(x - 1)$$

$$9e^{4x} - 4e^{2x} = e^{2x}(9e^{2x} - 4) = e^{2x}((3e^x)^2 - 2^2) = e^{2x}(3e^x + 2)(3e^x - 2)$$

$$(x + y)^2 - z^2 = \big((x + y) + z\big)\big((x + y) - z\big)$$

3. Sum/Difference of Cubes. These factoring techniques do not come very often, but when they do they can blindside the unprepared and make for an unpleasant day. They are:

$$a^3 + b^3 = (a + b)(a^2 - ab + b^2)$$

$$a^3 - b^3 = (a - b)(a^2 + ab + b^2)$$

Examples:
$$x^3 + 8 = x^3 + 2^3 = (x + 2)(x^2 - x \cdot 2 + 2^2) = (x + 2)(x^2 - 2x + 4)$$

$$e^{3x} - 27 = (e^x)^3 - 3^3 = (e^x - 3)((e^x)^2 + e^x \cdot 3 + 3^2) = (e^x - 3)(e^{2x} + 3e^x + 9)$$

$$125 - 8x^3 = 5^3 - (2x)^3 = (5 - 2x)(5^2 + 5 \cdot 2x + (2x)^2) = (5 - 2x)(25 + 10x + 4x^2)$$

4. Perfect Square Trinomials. These factoring principles are important to keep in mind, for they can be crucial to overcome an algebraic logjam. They are:

$$a^2 + 2ab + b^2 = (a + b)^2$$

$$a^2 - 2ab + b^2 = (a - b)^2$$

Think of these principles especially when you have a square root you would like to eliminate. For example, $\sqrt{x^2 + 10x + 25} = \sqrt{(x + 5)^2} = |x + 5|$ and $\sqrt{e^{2x} + 8e^x + 16} = \sqrt{(e^x + 4)^2} = e^x + 4$. One method for creating a perfect square trinomial is Completing the Square.

5. Completing the Square. There are two types of this factoring procedure: a.) the leading coefficient is equal to 1; b.) the leading coefficient is not equal to 1.

Examples:

$$x^2 + 4x + 7 \ = \ (x^2 + 4x + 2^2) + 7 - 2^2 \ = \ (x + 2)^2 + 7 - 4 \ = \ (x + 2)^2 + 3$$

$$2x^2 + 4x + 7 \ = \ 2(x^2 + 2x) + 7 \ = \ 2(x^2 + 2x + 1^2) + 7 - 2 \ = \ 2(x + 1)^2 + 5$$

There are two tricky aspects to the second type: Remember to factor out the value of the leading coefficient from the first *two* terms, and be careful about what quantity you need to subtract back to preserve equality.

2

The opposite process of factoring is expanding. *Expanding* is the process of multiplying out a factored expression into an equivalent series of terms. For example, $(x + 6)(x - 1)$ can be multiplied out by the "FOIL" method to $x^2 - x + 6x - 6$, which can be simplified to $x^2 + 5x - 6$. There are three additional expansion principles the calculus student should know: 2. Product of Binomial Conjugates; 3. Squared Binomials; 4. Cubed Binomials.

2. Product of Binomial Conjugates. This expansion method is the reverse of the factoring principle "Difference of Squares." Hence, in general, the pattern is

$$(a + b)(a - b) = a^2 - b^2$$

I suggest the student automatize this pattern without going through the intermediate labor of using the "FOIL" method. In words, the pattern would be: Write down the square of the first term, a^2, and subtract from that the square of the second term, $a^2 - b^2$.

Examples:

$$(x + 1)(x - 1) = x^2 - 1$$

$$(3x + 5)(3x - 5) = 9x^2 - 25$$

$$(4e^{2x} + 2e^{6x})(4e^{2x} - 2e^{6x}) = 16e^{4x} - 4e^{12x}$$

$$((x + y) + z)((x + y) - z) = (x + y)^2 - z^2$$

3. Squared Binomials. This expansion method is the opposite of the factoring principle Perfect Square Trinomials. The two general patterns of this expansion technique can be stated as

$$(a + b)^2 = a^2 + 2ab + b^2$$

or

$$(a - b)^2 = a^2 - 2ab + b^2$$

The student can write a squared binomial such as $(a + b)^2$ as a product of two binomials $(a + b)(a + b)$ and perform the "FOIL" method on the latter. However, I think a shorthand method is advisable in this case, since the squared binomial is a very common mathematical expression. For the first pattern, the shorthand method consists of taking the given squared binomial $(a + b)^2$ and writing out the square of the first term, a^2, adding it to *double* the product of the two terms, $a^2 + 2ab$, and then adding to the latter the square of the last term, $a^2 + 2ab + b^2$. The shorthand method for the second pattern is the same, except that one subtracts the double of the product of the two terms.

Examples:

$$(x + 1)^2 = x^2 + 2x + 1$$

$$(x - 3)^2 = x^2 - 6x + 9$$

$$(x^3 + 4)^2 = x^6 + 8x^3 + 16$$

$$(3e^{2x} - e^{5x})^2 = 9e^{4x} - 6e^{7x} + e^{10x}$$

$$\left((x + y) + z\right)^2 = (x + y)^2 + 2(x + y)z + z^2$$

4. Cubed Binomials. There is a principle called the "binomial formula" to immediately expand expressions of the form $(a + b)^3$ or $(a - b)^3$. However, I never found memorizing the binomial formula to be worth the time. Expanding cubed binomials is a rarity and there is an easy optional method: "FOIL" followed by multiplying by $(a \pm b)$. That is:

$$(a + b)^3 = (a + b)^2(a + b) = (a^2 + 2ab + b^2)(a + b) =$$

$$= a^3 + 2a^2b + ab^2 + a^2b + 2ab^2 + b^3 = a^3 + 3a^2b + 3ab^2 + b^3$$

and

$$(a - b)^3 = (a - b)^2(a - b) = (a^2 - 2ab + b^2)(a - b) =$$

$$= a^3 - 2a^2b + ab^2 - a^2b + 2ab^2 - b^3 = a^3 - 3a^2b + 3ab^2 - b^3$$

Summary Table of the Factoring and Expanding Methods

Factoring	Expanding
1. Greatest Common Factor	1. FOIL
2. Difference of Squares	2. Product of Binomial Conjugates
3. Sum/Difference of Cubes	3. Squared Binomials
4. Perfect Square Trinomials	4. Cubed Binomials
5. Completing the Square	

2. Fractions

Dealing with fractional expressions is very common in calculus. There are two broad topics regarding fractions that the new calculus student should be aware of: 1. the basic algebraic rules and manipulations; 2. Syntax options.

1. The basic algebraic rules and manipulations. The calculus student must have the basic algebra rules for fractions thoroughly automatized. (The rules discussed here pertain only to simple and complex fractions: "Mixed" numbers such as $4\frac{2}{3}$ do not arise in calculus.) To add two fractions, the denominators must be equal and one just adds the numerators, e.g., $\frac{2}{3} + \frac{x}{3} = \frac{2+x}{3}$. To multiply two fractions, one just multiples separately the numerators and denominators, e.g., $\frac{5}{8} \cdot \frac{x}{2} = \frac{5x}{16}$. If one wishes to add two fractions with unequal denominators, the fractions must be made to have identical denominators, e.g., $\frac{2}{3} + \frac{x}{4} = \frac{4}{4} \cdot \frac{2}{3} + \frac{x}{4} \cdot \frac{3}{3} = \frac{8}{12} + \frac{3x}{12} = \frac{8+3x}{12}$. To divide two fractions, multiply the top fraction by the reciprocal of the bottom fraction, e.g., $\frac{\frac{2x}{9}}{\frac{5}{7}} = \frac{2x}{9} \cdot \frac{7}{5} = \frac{14x}{45}$. To divide a fraction by an integer, such as $\frac{\left(\frac{4x}{5}\right)}{3}$, make the integer an explicitly equivalent fraction, viz., $\frac{\frac{4x}{5}}{\frac{3}{1}}$, and then proceed to compute, as $\frac{4x}{5} \cdot \frac{1}{3} = \frac{4x}{15}$. To divide an integer by a fraction, such as $\frac{3}{\left(\frac{4x}{5}\right)}$, make the integer an explicitly equivalent fraction, viz., $\frac{\frac{3}{1}}{\frac{4x}{5}}$, and then carry out the further computations, i.e., $\frac{3}{1} \cdot \frac{5}{4x} = \frac{15}{4x}$.

One situation that often arises in calculus is the need to reduce a complex fraction to a simple fraction. In arithmetic, a *simple fraction* is a rational number of the form $\frac{a}{b}$, where both a and b are integers ($b \neq 0$), e.g., $\frac{2}{3}$ and $\frac{8}{5}$. In algebra, a simple fraction can include examples with variables such as $\frac{2x}{3}$, where neither the numerator nor the denominator contains fractions. On the other hand, a *complex fraction* is a fraction whose numerator and/or denominator *does* contains fraction(s), e.g., $\frac{\left(\frac{7}{8}\right)}{15}$, $\frac{\frac{6}{5x}}{\frac{3}{8}}$, $\frac{\frac{3x}{5}+\frac{2}{3}}{6}$, $\frac{\frac{3}{5}+2}{\frac{5}{2}+\frac{22}{3x}}$. If the complex fraction contains no terms, as is the case for the first two examples immediately preceding, then just carry out the already-covered simplifications. If the complex fraction does contain terms, as is the case for the last two examples, then an additional step of adding fractions is necessary. For the example $\frac{\frac{3x}{5}+\frac{2}{3}}{6}$, we need to do the following: $\frac{\frac{3x}{5}+\frac{2}{3}}{6} = \frac{\frac{3}{3}\cdot\frac{3x}{5}+\frac{2}{3}\cdot\frac{5}{5}}{6} = \frac{\frac{9x}{15}+\frac{10}{15}}{6} = \frac{\frac{9x+10}{15}}{6} = \frac{\frac{9x+10}{15}}{\frac{6}{1}} = \frac{9x+10}{15} \cdot \frac{1}{6} = \frac{9x+10}{90}$. For the example $\frac{\frac{3}{5}+2}{\frac{5}{2}+\frac{22}{3x}}$, we do the following: $\frac{\frac{3}{5}+2}{\frac{5}{2}+\frac{22}{3x}} = \frac{\frac{3}{5}+2\cdot\frac{5}{5}}{\frac{3x}{3x}\cdot\frac{5}{2}+\frac{22}{3x}\cdot\frac{2}{2}} = \frac{\frac{3}{5}+\frac{10}{5}}{\frac{15x}{6x}+\frac{44}{6x}} = \frac{\frac{3+10}{5}}{\frac{15x+44}{6x}} = \frac{\frac{13}{5}}{\frac{15x+44}{6x}} = \frac{13}{5} \cdot \frac{6x}{15x+44} = \frac{78x}{5(15x+44)}$.

Finally, a manipulation that comes up occasionally in calculus has to do with multiplied fractions and the freedom to rearrange the order of the numerators or denominators. One example of this is $\frac{\sin(x)}{4\cos(x)} \cdot \frac{37}{x} = \frac{\sin(x)}{x} \cdot \frac{37}{4\cos(x)}$. The reason for changing the order of factors in the way I did has to do with the importance of grouping "$\frac{\sin(x)}{x}$" into its own fraction (which turns out to be in important expression in differential calculus).

2. Syntax options. Three important syntax issues related to fractions come up in calculus.

Firstly, there are two popular ways to write fractions, the first being "inline fractions," e.g., 2/3, and the second being "vertical fractions," e.g., $\frac{2}{3}$. I always recommend using vertical fractions, since it keeps a better visual distinction between the numerator and denominator. Furthermore, it helps in preventing a variable from morphing from being in the denominator to ending up in the numerator, e.g., 2/3x ... 2x/3.

Secondly, it is crucial in calculus to rewrite a fraction that looks like "$\frac{x^3}{2}$" to appear as "$\frac{1}{2}x^3$". Although both fractions are equivalent, the second way—where we are clearly demarcating the coefficient *to the left of the variable*—is indispensable in performing some of the differentiation rules.

Lastly, be sure to use enough parentheses when using calculators. A calculator will not know that "384 + 32.3/8" was meant to be the computation of (384 + 32.3)/8. It is better to use too many pairs of parentheses than too few!

3. Exponents and Radicals

Exponent principles play a major role in calculus. There are three fundamental such principles: 1. Same-Base Products; 2. Power of Powers; 3. Negative Exponents.

    1. Same-Base Products. Since $a^m \cdot a^n = a^{m+n}$, when multiplying numbers with the same real-number base, we add the exponents. (The exponents can be any real number.) Examples:

$$2^5 \cdot 2^3 = 2^{5+3} = 2^8.$$

$$3^{\frac{4}{5}} \cdot 3^{\frac{2}{5}} = 3^{\frac{4}{5}+\frac{2}{5}} = 3^{\frac{6}{5}}.$$

$$4^\pi \cdot 4^e = 4^{\pi+e}.$$

$$\pi^{\sqrt{2}} \cdot \pi^5 = \pi^{\sqrt{2}+5}.$$

$$x^7 \cdot x^{5\pi} = x^{7+5\pi}.$$

$$5^x \cdot 5^{\sin(\pi)} = 5^{x+\sin(\pi)} = 5^{x+0} = 5^x.$$

Furthermore, since $\dfrac{a^m}{a^n} = a^{m-n}$, when dividing numbers with the same real-number base, we subtract the exponents. (The exponents can be any real number.)

Examples:

$$\frac{2^5}{2^3} = 2^{5-3} = 2^2.$$

$$\frac{3^{\frac{4}{5}}}{3^{\frac{2}{5}}} = 3^{\frac{4}{5}-\frac{2}{5}} = 3^{\frac{2}{5}}.$$

$$\frac{4^\pi}{4^e} = 4^{\pi-e}.$$

$$\frac{\pi^{\sqrt{2}}}{\pi^5} = \pi^{\sqrt{2}-5}.$$

$$\frac{x^7}{x^{5\pi}} = x^{7-5\pi}.$$

$$\frac{5^x}{5^{\sin(\pi)}} = 5^{x-\sin(\pi)} = 5^{x-0} = 5^x.$$

7

2. Power of Powers. Since $(a^m)^n = a^{m \cdot n}$, when taking a power of a power, we multiply the exponents.

Examples:

$$(2^5)^3 = 2^{5 \cdot 3} = 2^{15}.$$ $$\left(3^{\frac{4}{5}}\right)^{\frac{2}{5}} = 3^{\frac{4}{5} \cdot \frac{2}{5}} = 3^{\frac{8}{25}}.$$

$$(4^\pi)^e = 4^{\pi \cdot e}.$$ $$\left(\pi^{\sqrt{2}}\right)^5 = \pi^{5\sqrt{2}}.$$

$$(x^7)^{5\pi} = x^{7 \cdot 5\pi} = x^{35\pi}.$$ $$(5^x)^{\sin(\pi)} = 5^{x \cdot \sin(\pi)} = 5^{x \cdot 0} = 5^0 = 1.$$

3. Negative Exponents. Since $a^{-n} = \frac{1}{a^n}$, and $\frac{1}{a^{-n}} = a^n$, when confronted by a base taken to a negative power, we can change the fractional location of the exponential and make the exponent positive.

Examples:

$$2^{-5} = \frac{1}{2^5}.$$ $$\frac{1}{3^{-\frac{4}{5}}} = 3^{\frac{4}{5}}.$$

$$4^{-\pi} = \frac{1}{4^\pi}.$$ $$\frac{1}{\pi^{-\sqrt{2}}} = \pi^{\sqrt{2}}.$$

$$x^{-7} = \frac{1}{x^7}.$$ $$\frac{1}{5^{-x}} = 5^x.$$

A crucial notational issue concerns the use of fractional exponents versus the use of radical signs. For example, $2^{\frac{1}{3}} = \sqrt[3]{2}$, $5^{\frac{4}{7}} = \sqrt[7]{5^4}$, and in general $a^{\frac{m}{n}} = \sqrt[n]{a^m}$. In calculus both notations are used heavily, depending on the context. When applying the derivative and integration rules, the exponential form is unavoidable. On the other hand, when engaging in algebraic manipulations to solve problems, the radical form is usually advantageous.

4. Algebraic Mal-Rules

Certain erroneous patterns crop up in algebra so often that they are given the special label "mal-rules."

The first type of pattern might be described as "false distribution." The first example of false distribution is where a student writes an equation such as

$$(a + b)^2 = a^2 + b^2.$$

This equation cannot be valid, since an easy counterexample can be constructed. Let $a = 1$ and $b = 1$. Now substitute these values into the left side of the pattern: $(1 + 1)^2 = 2^2 = 4$. Furthermore, substitute the same values into the right side of the same pattern: $1^2 + 1^2 = 1 + 1 = 2$. But $4 \neq 2$. Hence, the equation is not valid. The second example of false distribution is writing an equation of the form

$$\sqrt{a + b} = \sqrt{a} + \sqrt{b}.$$

Now this equation is also invalid, as shown by the following counterexample. Let $a = 16$ and $b = 9$. Substituting the values into the left side gives: $\sqrt{16 + 9} = \sqrt{25} = 5$. A similar substitution produces: $\sqrt{16} + \sqrt{9} = 4 + 3 = 7$. But $5 \neq 7$.

The second type of pattern might be termed "false cancellation." This common sort of error occurs when a student writes either

$$\frac{a + b}{a} = b$$

or

$$\frac{a + b}{a} = 1 + b$$

The first "equation" can be shown to be wrong by this counterexample: Let $a = 2$ and $b = 4$. $\frac{2+4}{2} = \frac{6}{2} = 3$. But $3 \neq 4$. The second's refutation can be shown by this counterexample: Let $a = 2$ and $b = 4$. $\frac{2+4}{2} = \frac{6}{2} = 3$. $1 + 4 = 5$. But $3 \neq 5$. Matters are not at all helped if the student writes

$$\frac{ac + b}{a} = c + b,$$

for the cancellation of the a's was illicit. (Let $a = 2$, $b = 2$, and $c = 3$.) The most common place where this latter false cancellation occurs is in the application of the quadratic formula.

Many a student is tempted to do something like the following:

$$x = \frac{4 \pm \sqrt{5}}{2} = 2 \pm \sqrt{5},$$

which is incorrect.

A third invalid pattern concerns a simple matter of mathematical punctuation, however young algebra students often overlook it, viz.,

$$(-a)^2 = -a^2.$$

Easy counterexample: Let $a = 2$. $(-2)^2 = (-2)(-2) = 4$. $-2^2 = -4$. But $4 \neq -4$.

A final mal-rule is more often a temptation than an actual commission, but it is a dangerous temptation nonetheless. This pattern may be termed "false fraction splitting," and looks like:

$$\frac{1}{a+b} = \frac{1}{a} + \frac{1}{b}.$$

A counterexample for this error is: Let $a = 2$ and $b = 4$. $\frac{1}{2+2} = \frac{1}{4}$. $\frac{1}{2} + \frac{1}{2} = 1$. But $\frac{1}{4} \neq 1$.

5. Difference Quotients

The difference quotient is a key component of the formal definition of the derivative. Most calculus textbooks state the general difference quotient as

$$\frac{f(x + h) - f(x)}{h},$$

where $f(x)$ is a function. For each function, there is a unique difference quotient. For example, the difference quotient of $f(x) = x^3$ is $\frac{(x-h)^3 - x^3}{h}$; the difference quotient of $f(x) = \sin(x)$ is $\frac{\sin(x+h) - \sin(x)}{h}$.

The task of a difference quotient problem is to find a way to algebraically resolve the original quotient so that the h in the original denominator can be cancelled out. There are three main types of functions that present different and challenging algebraic steps to resolve: 1. Expansion Type, e.g., $f(x) = x^2$; 2. Conjugate Type, e.g., $f(x) = \sqrt{x}$; 3. Fraction Type, e.g., $f(x) = \frac{1}{x}$. In essence, there are just three steps to solve these problems: *Write out, substitute,* and *resolve.*

1. Expansion Type. The simplest function of this type is $f(x) = x^2$. To resolve its difference quotient, perform the following steps:

a. <u>Write out $f(x)$ and $f(x + h)$</u>: $f(x) = x^2 \quad f(x + h) = (x + h)^2$.

b. <u>Write out the general difference quotient and then substitute for $f(x)$ and $f(x + h)$ what you just wrote out above on the right sides of the equals sign:</u>

$$\frac{f(x + h) - f(x)}{h} = \frac{(x + h)^2 - x^2}{h} = \cdots$$

c. <u>Resolve the difference quotient to be able to cancel out the denominator's h:</u>

$$\cdots = \frac{x^2 + 2xh + h^2 - x^2}{h} = \frac{2xh - h^2}{h} = \frac{h(2x - h)}{h} = 2x - h.$$

Therefore: $\frac{f(x+h) - f(x)}{h} = 2x - h$.

2. Conjugate Type. The simplest function of this type is $f(x) = \sqrt{x}$. To resolve its difference quotient, perform the following steps:

a. Write out: $f(x) = \sqrt{x}$ $f(x+h) = \sqrt{x+h}$.

b. Substitute:

$$\frac{f(x+h) - f(x)}{h} = \frac{\sqrt{x+h} - \sqrt{x}}{h} = \cdots$$

c. Resolve:

$$\cdots = \frac{\sqrt{x+h} - \sqrt{x}}{h} \cdot \frac{\sqrt{x+h} + \sqrt{x}}{\sqrt{x+h} + \sqrt{x}} = \frac{(x+h) - x}{h(\sqrt{x+h} + \sqrt{x})} = \frac{h}{h(\sqrt{x+h} + \sqrt{x})} = \frac{1}{(\sqrt{x+h} + \sqrt{x})}.$$

Therefore: $\frac{f(x+h) - f(x)}{h} = \frac{1}{\sqrt{x+h} + \sqrt{x}}$.

(Note: The h that now exists as a term inside the radical sign is okay—as you will find out early in calculus.)

3. Fraction Type. The simplest function of this type is $f(x) = \frac{1}{x}$. To resolve its difference quotient, perform the following steps:

a. Write out: $f(x) = \frac{1}{x}$ $f(x+h) = \frac{1}{x+h}$.

b. Substitute:

$$\frac{f(x+h) - f(x)}{h} = \frac{\dfrac{1}{x+h} - \dfrac{1}{x}}{h} = \cdots$$

c. Resolve:

$$\cdots = \frac{\dfrac{x}{x} \cdot \dfrac{1}{x+h} - \dfrac{1}{x} \cdot \dfrac{x+h}{x+h}}{h} = \frac{\dfrac{x}{x(x+h)} - \dfrac{(x+h)}{x(x+h)}}{h} = \frac{\dfrac{x - (x+h)}{x(x+h)}}{h} = \frac{\dfrac{x - x - h}{x(x+h)}}{h} = \frac{\dfrac{-h}{x(x+h)}}{h} =$$

$$= \frac{\dfrac{-h}{x(x+h)}}{\dfrac{h}{1}} = \frac{-h}{x(x+h)} \cdot \frac{1}{h} = \frac{-1}{x(x+h)}.$$

Therefore, $\frac{f(x+h) - f(x)}{h} = \frac{-1}{x(x+h)}$.

A challenging combination of the aforementioned types is a mixture of types 2 and 3, e.g., the function $f(x) = \frac{1}{\sqrt{x}}$. To resolve its difference quotient, perform the following steps:

a. <u>Write out</u>: $f(x) = \frac{1}{\sqrt{x}}$ $f(x+h) = \frac{1}{\sqrt{x+h}}$.

b. <u>Substitute</u>:

$$\frac{f(x+h) - f(x)}{h} = \frac{\frac{1}{\sqrt{x+h}} - \frac{1}{\sqrt{x}}}{h} = \cdots$$

c. <u>Resolve</u>:

$$\cdots = \frac{\frac{\sqrt{x}}{\sqrt{x}} \cdot \frac{1}{\sqrt{x+h}} - \frac{1}{\sqrt{x}} \cdot \frac{\sqrt{x+h}}{\sqrt{x+h}}}{h} = \frac{\frac{\sqrt{x}}{\sqrt{x}\sqrt{x+h}} - \frac{\sqrt{x+h}}{\sqrt{x}\sqrt{x+h}}}{h} = \frac{\frac{\sqrt{x} - \sqrt{x+h}}{\sqrt{x}\sqrt{x+h}}}{h} =$$

$$= \frac{\frac{\sqrt{x} - \sqrt{x+h}}{\sqrt{x}\sqrt{x+h}} \cdot \frac{\sqrt{x} + \sqrt{x+h}}{\sqrt{x} + \sqrt{x+h}}}{h} = \frac{\frac{x - (x+h)}{\sqrt{x}\sqrt{x+h}(\sqrt{x} + \sqrt{x+h})}}{h} = \frac{\frac{x - x - h}{\sqrt{x}\sqrt{x+h}(\sqrt{x} + \sqrt{x+h})}}{h} =$$

$$= \frac{\frac{-h}{\sqrt{x}\sqrt{x+h}(\sqrt{x} + \sqrt{x+h})}}{h} = \frac{\frac{-h}{\sqrt{x}\sqrt{x+h}(\sqrt{x} + \sqrt{x+h})}}{\frac{h}{1}} = \frac{-h}{\sqrt{x}\sqrt{x+h}(\sqrt{x} + \sqrt{x+h})} \cdot \frac{1}{h} =$$

$$= \frac{-1}{\sqrt{x}\sqrt{x+h}(\sqrt{x} + \sqrt{x+h})}.$$

Therefore: $\frac{f(x+h)-f(x)}{h} = \frac{-1}{\sqrt{x}\sqrt{x+h}(\sqrt{x}+\sqrt{x+h})}$.

13

A final challenging variation looks similar to a Conjugate Type. It is $(x) = \sqrt[3]{x}$. However, to resolve it requires an appeal to the difference of cubes formula from algebra,

$$a^3 - b^3 = (a - b)(a^2 + ab + b^2).$$

Here is a procedure for resolving this function's difference quotient, where $a = \sqrt[3]{x+h}$ and $b = \sqrt[3]{x}$.

a. <u>Write out</u>: $f(x) = \sqrt[3]{x}$ $\quad f(x + h) = \sqrt[3]{x+h}$.

b. <u>Substitute</u>:

$$\frac{f(x + h) - f(x)}{h} = \frac{\sqrt[3]{x+h} - \sqrt[3]{x}}{h} = \cdots$$

c. <u>Resolve</u>:

$$\cdots = \frac{\sqrt[3]{x+h} - \sqrt[3]{x}}{h} \cdot \frac{\sqrt[3]{(x+h)^2} + \sqrt[3]{x+h}\sqrt[3]{x} + \sqrt[3]{x^2}}{\sqrt[3]{(x+h)^2} + \sqrt[3]{x+h}\sqrt[3]{x} + \sqrt[3]{x^2}} =$$

$$= \frac{\sqrt[3]{(x+h)^3} + \sqrt[3]{(x+h)^2}\sqrt[3]{x} + \sqrt[3]{x+h}\sqrt[3]{x^2} - \sqrt[3]{(x+h)^2}\sqrt[3]{x} - \sqrt[3]{x+h}\sqrt[3]{x^2} - \sqrt[3]{x^3}}{h\left(\sqrt[3]{(x+h)^2} + \sqrt[3]{x+h}\sqrt[3]{x} + \sqrt[3]{x^2}\right)} =$$

$$= \frac{(x+h) + \sqrt[3]{(x+h)^2}\sqrt[3]{x} - \sqrt[3]{(x+h)^2}\sqrt[3]{x} + \sqrt[3]{x+h}\sqrt[3]{x^2} - \sqrt[3]{x+h}\sqrt[3]{x^2} - x}{h\left(\sqrt[3]{(x+h)^2} + \sqrt[3]{x+h}\sqrt[3]{x} + \sqrt[3]{x^2}\right)} =$$

$$= \frac{x + h - x}{h\left(\sqrt[3]{(x+h)^2} + \sqrt[3]{x+h}\sqrt[3]{x} + \sqrt[3]{x^2}\right)} = \frac{h}{h\left(\sqrt[3]{(x+h)^2} + \sqrt[3]{x+h}\sqrt[3]{x} + \sqrt[3]{x^2}\right)} =$$

$$= \frac{1}{\sqrt[3]{(x+h)^2} + \sqrt[3]{x+h}\sqrt[3]{x} + \sqrt[3]{x^2}} \cdot$$

Therefore: $\dfrac{f(x+h)-f(x)}{h} = \dfrac{1}{\sqrt[3]{(x+h)^2}+\sqrt[3]{x+h}\sqrt[3]{x}+\sqrt[3]{x^2}} \cdot$

6. Functions

Functions are a central issue in calculus. Three issues about functions that need attention are: 1. the general definition of the function concept; 2. the domain and range of a function; 3. the understanding of function notation.

1. The general definition of the function concept. A simple way to understand what functions do is to imagine a "rule machine" that takes in inputs and produces outputs. Let x represent the inputs and y the outputs. Now this machine produces the outputs according to a rule specified by an equation. For example, one rule machine could be:

$$x \Longrightarrow \boxed{y = 2x + 3} \Longrightarrow y$$

In words, we could say that the rule machine takes an input, multiplies it by 2, and then adds 3 to the product. Some examples of inputs becoming outputs would be:

$$1 \Longrightarrow \boxed{y = 2(1) + 3} \Longrightarrow 5$$

$$2 \Longrightarrow \boxed{y = 2(2) + 3} \Longrightarrow 7$$

$$3 \Longrightarrow \boxed{y = 2(3) + 3} \Longrightarrow 9$$

Typically the inputs and outputs are communicated as *ordered pairs* of the form (x, y). Hence, for the examples about we have $(1, 5)$, $(2, 7)$, and $(3, 9)$.

The only restriction on the function rule machine is that each input produces a *unique* output. When graphing functions, this restriction is often described as the "vertical line test." For example, the equation $y = x^2$ is a function, since it passes the vertical line test. (See the left graph below.) However, the equation $x = y^2$ is not a function, since it fails the same test. (See the right graph below.)

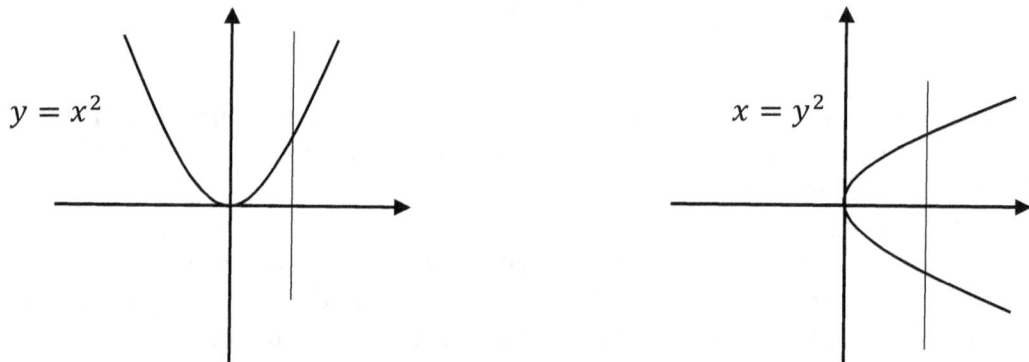

A last issue regarding the basic concept of functions concerns the letters standing for the inputs and outputs. Let us adopt the shorthand $x \to y$ to stand in place of the rule machine diagrams we used above. With that shorthand, the student will often see $t \to P$, where t (for time) is the input and P is the output. Other popular choices are $\theta \to y$ (often used in trigonometric applications) and $t \to x$.

2. The domain and range of a function. In addition to understanding the basic function concept, the calculus student will face determining the domain and range of a given function. In terms of the rule machine metaphor, the domain of a function is just the set of all the inputs; the range is just the set of all the outputs. Nothing more, nothing less.

Often the domain and range will have to be determined from the function's graph. Here it is important to grasp the principle that the domain is found by tracking the graph relative to the x-axis going left to right. Whatever x values are included in the graph constitute the domain of the function. In a similar manner, the range is found by tracking the graph relative to the y-axis going from the bottom to the top. Whatever y values are included in the graph constitute the range of the function.

3. The understanding of function notation. The ordered pair (x, y) can be written using function notation, which consists of an equation such as $f(x) = y$. Using this equation, we can identify the three parts to this notation. From left to right, the first part is the name of the function. It can be a single letter or a whole word. The second part is the pair of parentheses, which contain the input(s). Each input is called an "argument" of the function. Finally, the third part—on the right side of the equality sign—is the output corresponding to the given input. Each such output is called a "value" of the function.

$$f(x) = y$$

name   argument   value

Returning to the three ordered pairs used in the rule machine, we would have $f(1) = 5$, $f(2) = 7$, and $f(3) = 9$.

16

7. Graphs

There are six basic graphs that are worth memorizing for calculus:

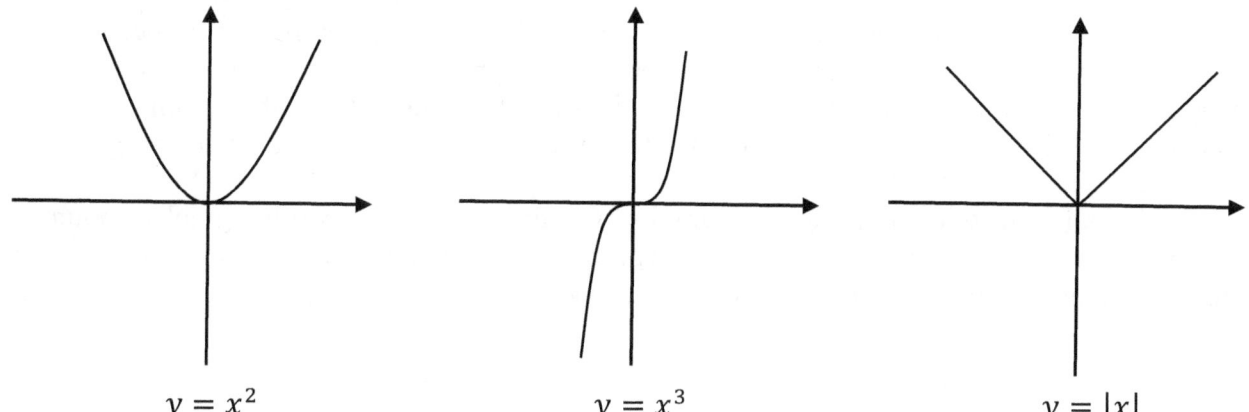

$$y = x^2 \qquad\qquad y = x^3 \qquad\qquad y = |x|$$

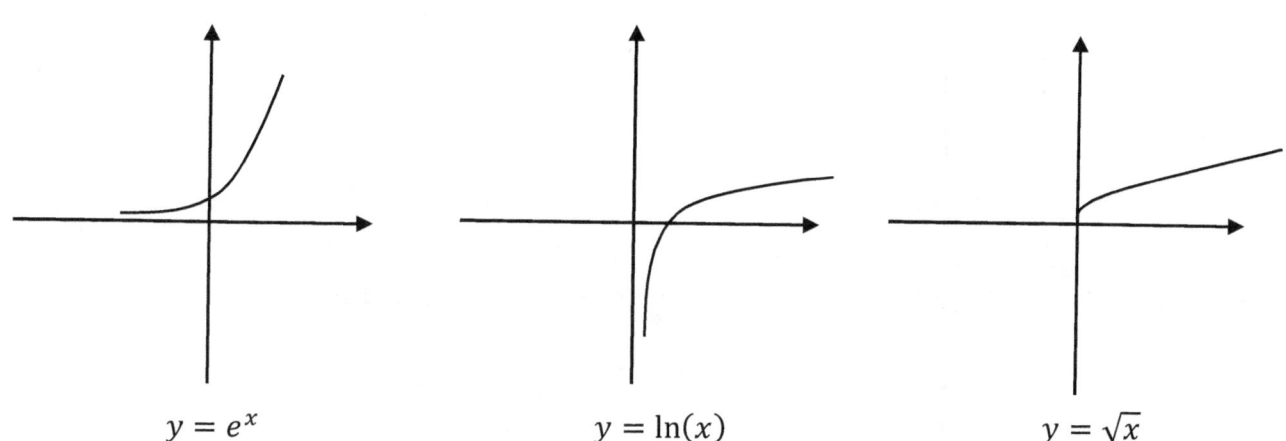

$$y = e^x \qquad\qquad y = \ln(x) \qquad\qquad y = \sqrt{x}$$

With the graphs memorized, the student can answer questions about the domain and range of the graphed functions. For example, by inspecting the graph for $y = e^x$, one can go left to right and conclude that the domain includes all x values, which in interval notation would be expressed $(-\infty, \infty)$. Furthermore, by tracking from the bottom up, one can also determine the range to be all y values greater than 0, i.e., $(0, \infty)$. A further virtue of memorizing the graph for $y = e^x$ is that one can see that as x goes off to the left in the negative direction, the function's y-values approach ever closer to 0. That is, e^x has a horizontal asymptote $y = 0$.

A second value of memorizing these graphs is that more involved functions such as $y = x^3 + 3x^2 + 5x + 10$ follow the pattern of the basic graph $y = x^3$ with large values of x (whether positive or negative). The reason for this is that the leading term of the complex function—in this case x^3—tends to predominate in influence as the absolute value of x grows very large.

A final value of committing these graphs to memory is that many other graphs amount to a set of shiftings and/or reflections of the basic graphs. For example, the graph of $y = (x - 3)^2$ can be thought of as the graph of the basic $y = x^2$ moved to the right three units:

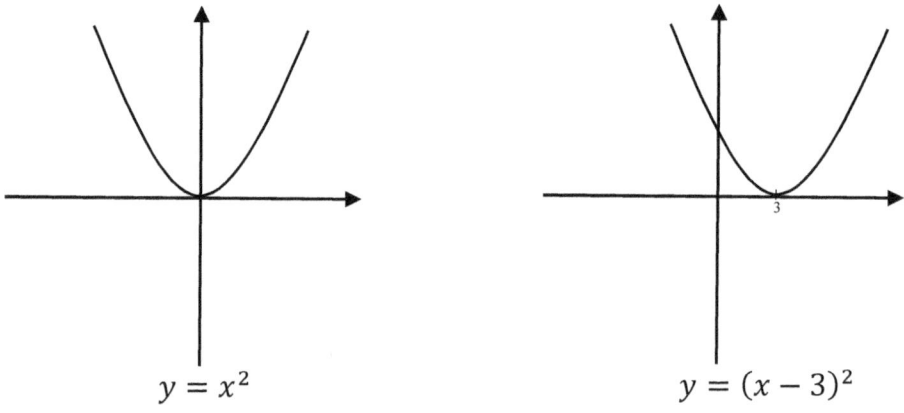

$$y = x^2 \qquad\qquad\qquad y = (x-3)^2$$

In a different case, the graph of $y = |x| + 4$ can be regarded as the basic graph of $y = |x|$ moved up four units:

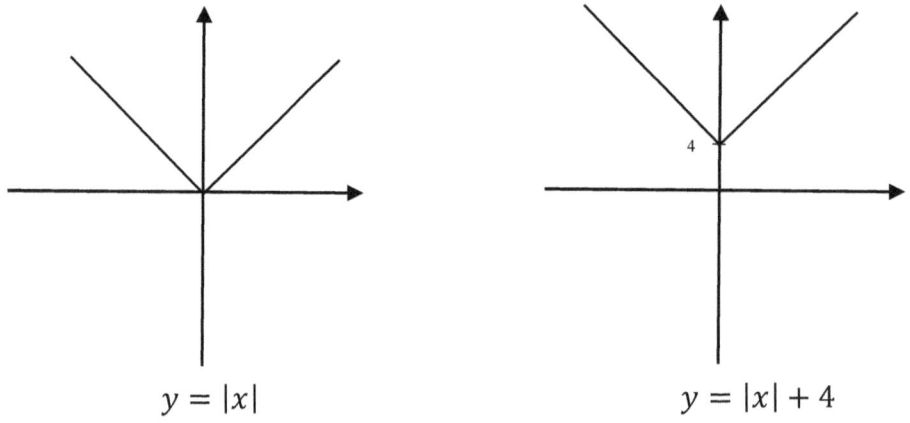

$$y = |x| \qquad\qquad\qquad y = |x| + 4$$

8. Linear Functions

Linear functions act as a sort of foundation for calculus, both theoretically and in applications. Hence, mastering the principles of such functions is crucial for the calculus student.

To begin, a linear function is called that because its graph is a straight line. For graphing purposes, the best type of equation for a linear function is called the slope-intercept form, usually abbreviated by

$$y = mx + b,$$

Where m represents the slope and b the y-intercept.

Recall that the y-intercept always has its x-value equal to zero. In other words, the y-intercept b corresponds to the point $(0, b)$:

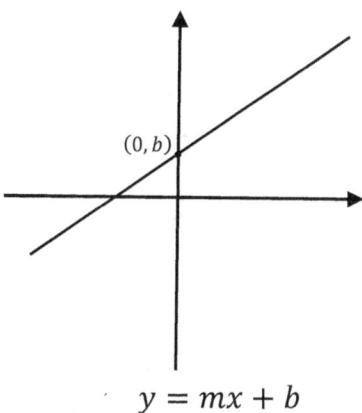

$$y = mx + b$$

In calculus, the slope of a linear function is often called the "average rate of change." In either case, the formula for finding the slope is "rise over run," also known as "delta y over delta x" or "the change in y over the change in x":

$$m = \frac{y_2 - y_1}{x_2 - x_1} \, .$$

A common type of problem involving linear functions is where a point of the function is given as well as the slope, and the student is requested to write out the equation of the line. For example, it might be given that $(3, 2)$ is a point on the line and the slope is $m = 5$. There are two popular ways to complete this equation, one using the point-slope form of the equation and the other going directly to the slope-intercept form.

1. Using the point-slope form. The point slope form is:

$$y_2 - y_1 = m(x_2 - x_1).$$

Hence, by substitution, we have

$$y - 2 = 5(x - 3)$$

and solving for y, we get

$$y - 2 = 5x - 15$$

$$y = 5x - 13$$

2. Using the slope-intercept form. The slope-intercept form is:

$$y = mx + b.$$

Hence, by substitution, we have

$$2 = 5(3) + b$$

And solving for b, we get

$$2 = 15 + b$$

$$-13 = b.$$

So,

$$y = 5x - 13.$$

Finally, the calculus student must automatically associate a positive slope with the straight line rising from left to right, a negative slope with the straight line falling from left to right, and zero slope with a horizontal line. For example:

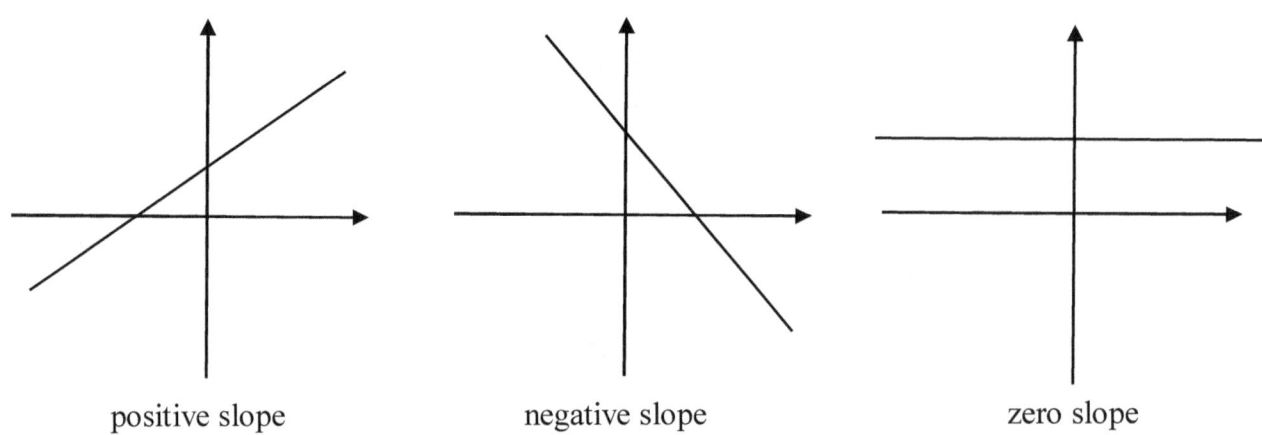

positive slope   negative slope   zero slope

9. Plane Geometry

The geometry a calculus student needs falls into three categories: 1. Basic formulas; 2. Triangle principles; 3. Drawing figures.

1. The following table represents the maximum number of geometry formulas that the calculus student may be expected to have memorized.

A – Area C – Circumference V – Volume SA – Surface Area P – Perimeter
r – radius h – height b – base s – side l – length w – width

Circles:	$C = 2\pi r$	$A = \pi r^2$
Cylinders:	$SA = 2\pi r^2 + 2\pi rh$	$V = \pi r^2 h$
Cones:	$V = \frac{1}{3}\pi r^2 h$	
Spheres:	$A = 4\pi r^2$	$V = \frac{4}{3}\pi r^3$
Triangles:	$A = \frac{1}{2}bh$	
Rectangles:	$P = 2l + 2w$	$A = lw$
Squares:	$P = 4s$	$A = s^2$
Rectangular Prisms:	$SA = 2wh + 2wl + 2hl$	$V = lwh$
Cubes:	$SA = 6s^2$	$V = s^3$

The above formulas for spheres are usually the most difficult for students to memorize. As an aid, when you learn about the differentiation formula called the Power Rule, note that the derivative of the volume of a sphere equals the area of the sphere. (Also, the derivative of a circle's area equals the circumference.)

2. Three principles of triangles are of utmost importance in calculus: the Pythagorean Theorem, properties of similar triangles, and the construction of the two reference triangles.

The Pythagorean Theorem, e.g.,

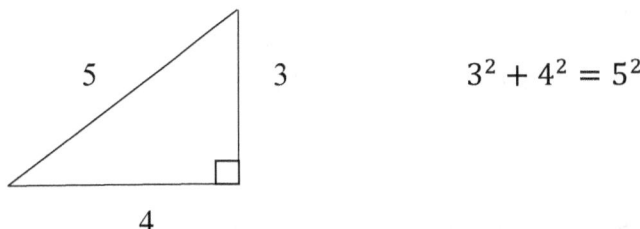

$$3^2 + 4^2 = 5^2$$

is so crucial, that anytime a right triangle comes up in a problem, it is very likely that the theorem is needed to solve it.

Similar triangles come up occasionally in calculus, especially in word problems. The salient property of similar triangles is that their corresponding sides are proportional. For example, note that the corresponding sides of the following two similar triangles equal the same ratio $\frac{1}{2}$.

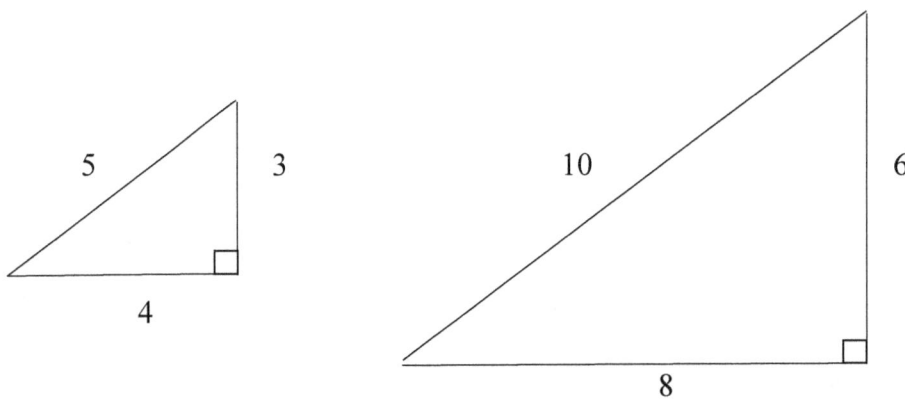

3. The two reference triangles are the 45°-45°-90° and 30°-60°-90° triangles. These triangles are some of the most important tools in mathematics, having applications in geometry, trigonometry, calculus, and word problems.

The best way to construct the 45°-45°-90° reference triangle is to let the legs equal 1 and (by the Pythagorean Theorem) derive the hypotenuse to equal $\sqrt{2}$.

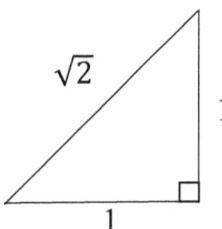

This first reference triangle is relatively easy to construct, and students almost always do so correctly. However, the second reference triangle–30°-60°-90°–is more prone to error.

To construct a correct 30°-60°-90° reference triangle, do the following steps:

1. Draw an equilateral triangle with side lengths 2. Label the fact that all the angles are 60°.

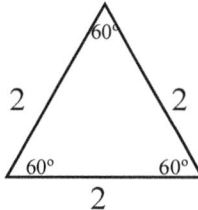

2. Drop down a perpendicular bisector from the top angle to the bottom side.

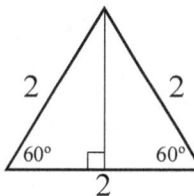

3. Focus on the left right triangle. Label the fact that the shorter leg is 1 and the top angle is 30°.

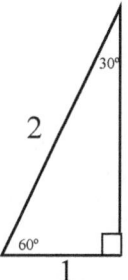

23

4. Use the Pythagorean Theorem to find the longer leg's length:

$$1^2 + b^2 = 2^2 \quad \Longrightarrow \quad b^2 = 4 - 1 \quad \Longrightarrow \quad b = \sqrt{3}$$

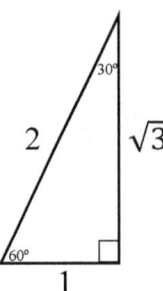

Note the fact that the hypotenuse is double the length of the shorter leg. Note further that the longer leg is $\sqrt{3}$ times the length of the shorter leg.

3. Drawing geometric figures. Drawing reasonable looking three-dimensional figures helps us to solve application problems.

Drawing a sphere comes down to drawing a circle with a horizontal oval inside:

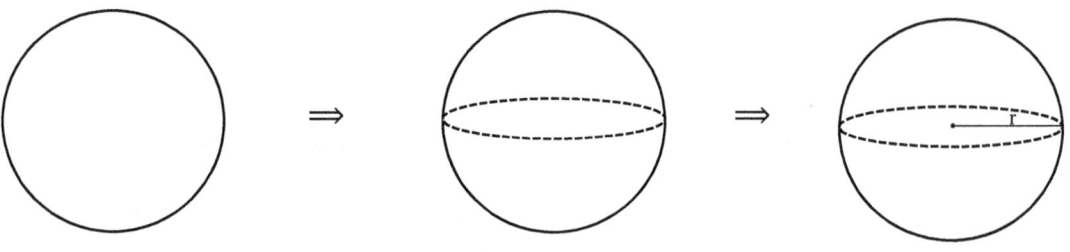

Drawing a cone comes down to drawing an isosceles triangle on top of a horizontally-oriented oval:

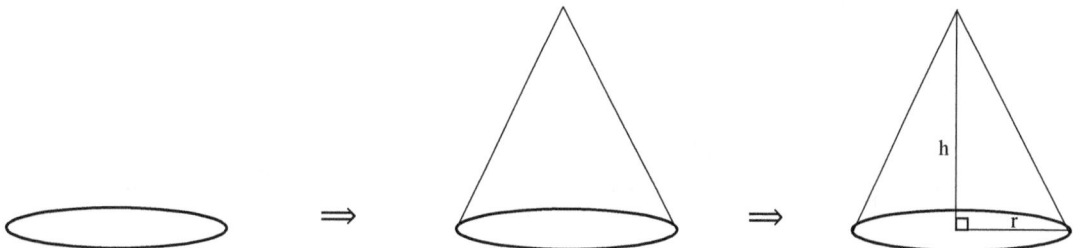

Drawing a vertical cylinder amounts to drawing two connected horizontally-oriented ovals, one directly above the other:

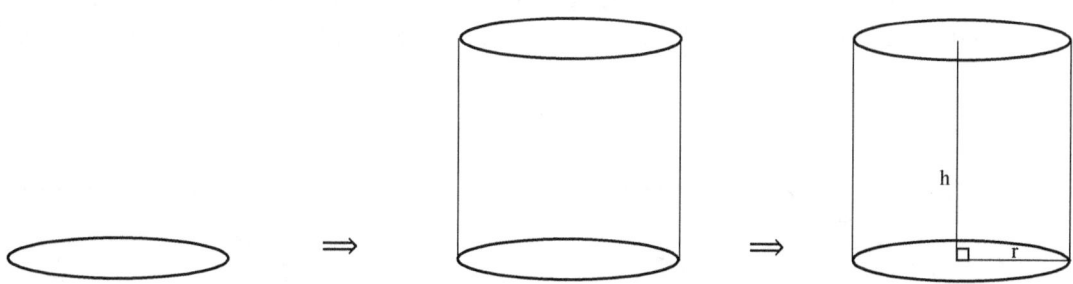

Whereas drawing a horizontal cylinder is a matter of drawing two connected circles, one "behind" the other:

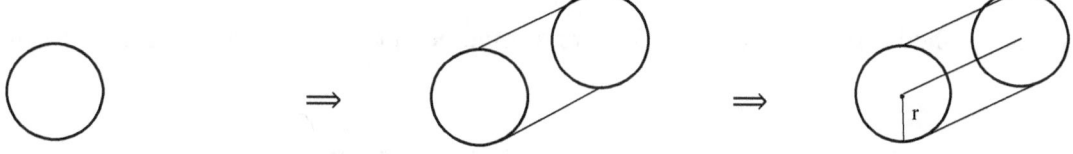

To draw a rectangular box, first draw a rectangle. Next draw lines that start at the two top corners, go up to the right, and *are parallel to each other*. Make sure that those two lines are the same length and that a horizontal line can join their ends.

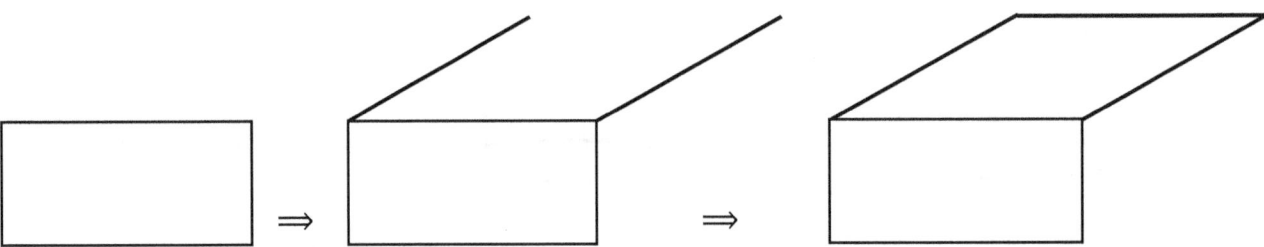

Now from the lower right corner of the rectangle draw a line that is parallel to the top lines. Draw it long enough so that a vertical line can join the upper right corner of the top to the end of the line.

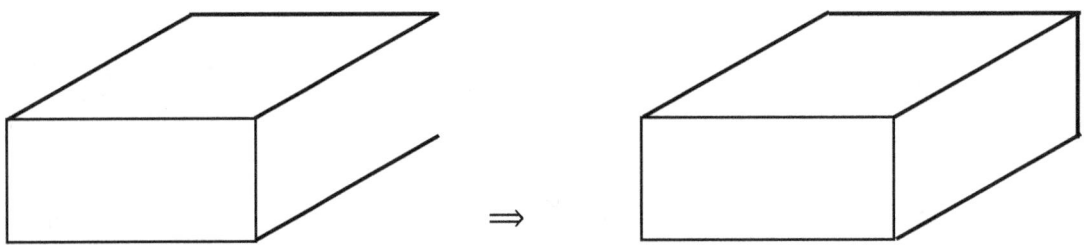

Finally, add dashed lines parallel to the others to give perspective to the "unseen" part of the box.

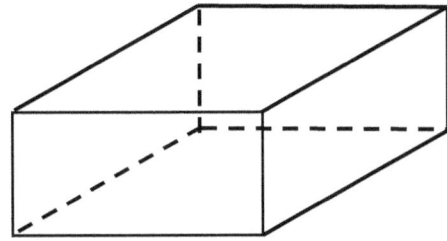

The key principle in drawing the box is parallel lines. Note carefully the set of horizontal lines, the set of vertical lines, and the set of horizontal lines that shoot off to the upper right. Each set has four lines.

10. Trigonometry

The needed trigonometric knowledge a successful calculus student must have falls into five categories: 1. The basis of the trigonometric functions on similar right triangles; 2. The domain versus range of trigonometric functions with respect to similar right triangles and the unit circle; 3. Needed trigonometric identities; 4. Radian measure; 5. Unit circle measurements.

1. Similar right triangles. Consider two similar right triangles.

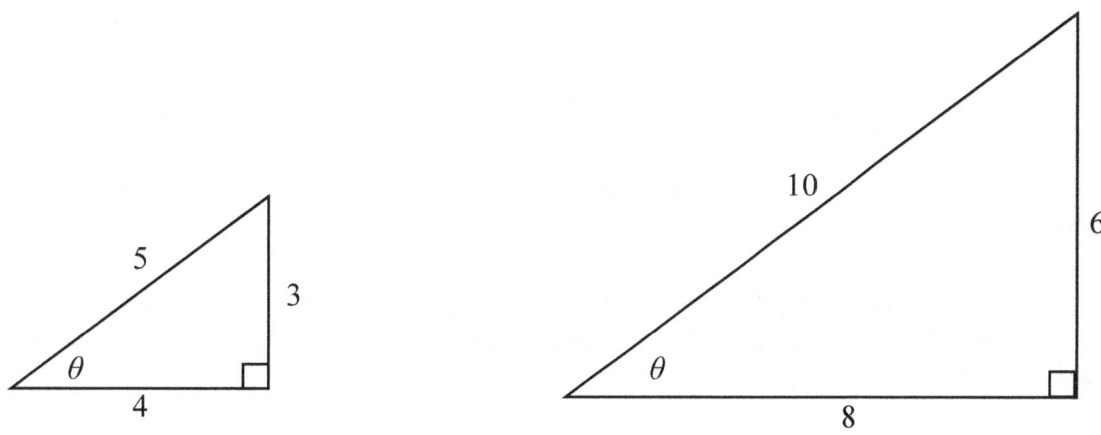

The angle θ is the same for both triangles. In the left triangle, the leg opposite to θ has length 3 and the leg adjacent to θ has length 4. Hence, the ratio of opposite over adjacent equals $\frac{3}{4}$. Notice that in the case of the triangle on the right the ratio of opposite leg to θ over the adjacent leg to θ equals $\frac{6}{8} = \frac{3}{4}$. Pick any other similar triangles to these first two and the same ratio of $\frac{3}{4}$ appears for the leg opposite to θ divided by leg adjacent to θ. For brevity, let us abbreviate the ratio as $\frac{\text{OPP}}{\text{ADJ}}$. The ratio $\frac{\text{OPP}}{\text{ADJ}}$ is called "tangent." That is, the mathematical tangent of $= \frac{\text{OPP}}{\text{ADJ}}$, with the shorthand $\tan(\theta) = \frac{\text{OPP}}{\text{ADJ}}$. Note carefully that the tangent of angle θ *is* the ratio of the opposite leg divided by the adjacent leg.

Furthermore, in all the above triangles, the ratio of the leg opposite θ divided by the hypotenuse equals $\frac{3}{5}$, viz., $\frac{\text{OPP}}{\text{HYP}} = \frac{3}{5}$. The trigonometric function that specifies the ratio $\frac{\text{OPP}}{\text{HYP}}$ relative to θ is called "sine." More briefly, $\sin(\theta) = \frac{\text{OPP}}{\text{HYP}}$. Lastly, "cosine" is defined as $\cos(\theta) = \frac{\text{ADJ}}{\text{HYP}}$.

It is crucial to remember that sine, cosine, and tangent are defined *relative to an angle*. For example, in the following triangle

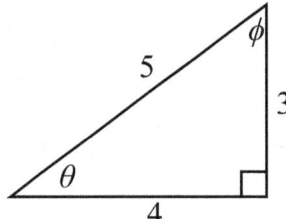

$\sin(\theta) = \frac{3}{5}$, but $\sin(\phi) = \frac{4}{5}$; $\tan(\theta) = \frac{3}{4}$, but $\tan(\phi) = \frac{4}{3}$. Note however that $\cos(\theta) = \frac{4}{5} = \sin(\phi)$. That is, the *co*sine of θ is equal to the sine of ϕ. But $\theta + \phi = 90°$, that is, θ and ϕ are *complementary* angles. Hence, "cosine" is short for "complementary angle's sine."

2. Domain and range. When dealing with right triangles, the domain of the sine, cosine, and tangent functions consists of *acute angles*. The range of the same functions consists of ratios of side lengths. For example, in the following triangle:

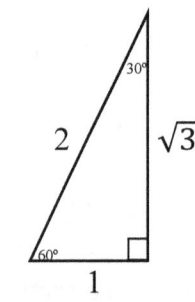

we have

$$\sin(30°) = \frac{1}{2}$$

A member of the domain   A member of the range

When the trigonometric functions are applied to the unit circle, the domain is expanded to include all real numbers and the range is the interval $[-1, 1]$.

3. Needed identities. For sure a calculus student must have memorized the following trigonometric identities:

$$\tan(\theta) = \frac{\sin(\theta)}{\cos(\theta)} \qquad \sec(\theta) = \frac{1}{\cos(\theta)} \qquad \csc(\theta) = \frac{1}{\sin(\theta)} \qquad \cot(\theta) = \frac{1}{\tan(\theta)}$$

Furthermore, the Pythagorean identity must be automatized:

$$\boxed{\sin^2(\theta) + \cos^2(\theta) = 1}$$

There are two variations of the Pythagorean identity that are especially important in integral calculus. They can easily be derived on the fly as follows.

$$\sin^2(\theta) + \cos^2(\theta) = 1 \qquad\qquad \text{State the Pythagorean identity.}$$

$$\frac{\sin^2(\theta)}{\cos^2(\theta)} + \frac{\cos^2(\theta)}{\cos^2(\theta)} = \frac{1}{\cos^2(\theta)} \qquad\qquad \text{Divide both sides by } \cos^2(\theta).$$

$$\boxed{\tan^2(\theta) + 1 = \sec^2(\theta)} \qquad\qquad \text{Simplify.}$$

The second variation follows by a similar line of reasoning.

$$\sin^2(\theta) + \cos^2(\theta) = 1 \qquad\qquad \text{State the Pythagorean identity.}$$

$$\frac{\sin^2(\theta)}{\sin^2(\theta)} + \frac{\cos^2(\theta)}{\sin^2(\theta)} = \frac{1}{\sin^2(\theta)} \qquad\qquad \text{Divide both sides by } \sin^2(\theta).$$

$$\boxed{1 + \cot^2(\theta) = \csc^2(\theta)} \qquad\qquad \text{Simplify.}$$

4. Radian measure. When thinking of trigonometric functions as applied to circles, calculus uses only radian measure.

Consider a circle with central angle θ with a radius of four inches.

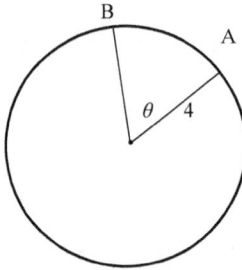

If the arc length of $\overset{\frown}{AB}$ also equals four inches,

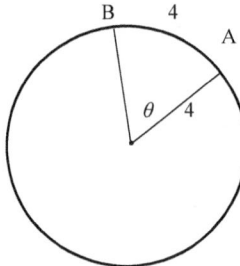

then θ is said to have a radian measure of one. If instead the arc length of $\overset{\frown}{AB}$ equals eight while the radius remains at four,

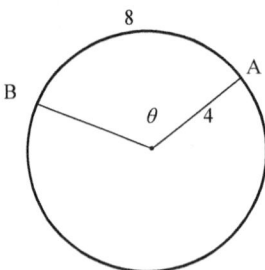

then θ is said to have a radian measure of two. In other words, the radian measure begins by marking off arc lengths equal to the radius of the circle.

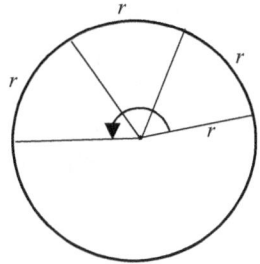

The central angle's radian measure here is three.

To generalize: The radian measure of a central angle subtended by an arc is the number of radii included in the arc. Or: The radian measure of a central angle is the number of radii included in the arc intercepted by that angle. Symbolically, for arc length s and radius r, the radian measure of central angle θ is defined by

$$\theta = \frac{s}{r}.$$

For example, for an arc of length eight that subtends angle θ in a circle of radius five means that θ measures $\frac{8}{5}$ radians.

Now consider the case where the arc length involved is the entire circumference C of the circle, viz.,

$$\theta = \frac{s}{r} = \frac{C}{r} = \frac{2\pi r}{r} = 2\pi.$$

Hence the radian measure of a central angle that involves a complete rotation equals 2π radians. So, $360° = 2\pi$ radians. (In calculus, if an angle does not have the degree mark " ° ", then mathematicians assume that the angle is measured in radians. Thus, we may write simply $360° = 2\pi$.)

The fact that $360° = 2\pi$ implies that $180° = \pi$ allows us to state that $\frac{180°}{180} = 1° = \frac{\pi}{180}$, i.e., one degree equals $\frac{\pi}{180}$ radians. This equation gives us a way to convert from degrees to radians. For example, since $1° = \frac{\pi}{180}$, $30° = \frac{30\pi}{180} = \frac{\pi}{6}$. Furthermore, since $180° = \pi$, we can deduce that $\frac{180°}{\pi} = \frac{\pi}{\pi} = 1$, i.e., one radian equals $\frac{180}{\pi}$ degrees. This latter equation will let us convert from radians to degrees, e.g., since $\frac{180°}{\pi} = 1$, $\frac{180°}{\pi} \cdot \frac{\pi}{3} = 60° = \frac{\pi}{3}$.

5. Unit circle. Many a high school student is forced to memorize the *xy* coordinates on the unit circle. An easier way to compute these values (and more) is by the use of "reference angles."

Let θ be an angle in standard position that is not a multiple of 90°. Then the reference angle for angle θ placed in standard position is the angle θ', which is the positive acute angle formed by the terminal side of θ and the horizontal axis. Examples:

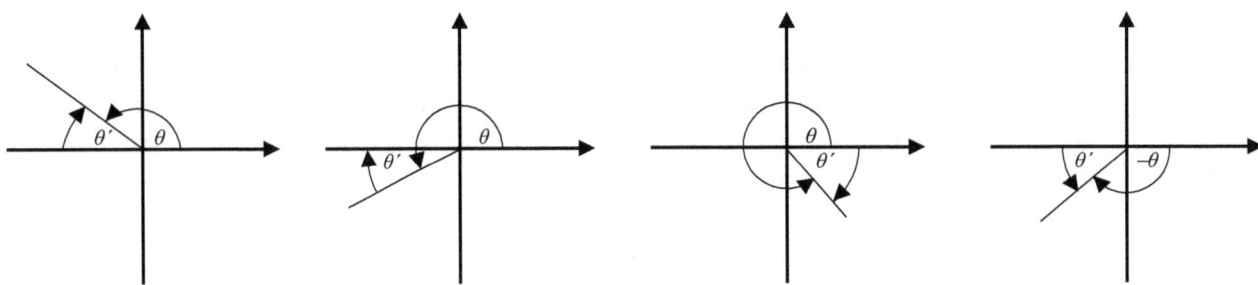

(Important note: Reference angles *never* "come out" of the vertical axis.) A concrete example is the following: the reference angle of a 120° angle is 60°.

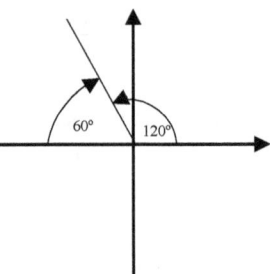

The reference angle of an angle of $\frac{11\pi}{6}$ is $\frac{\pi}{6}$:

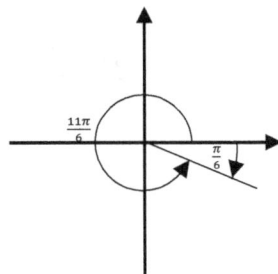

Now $\sin\left(\frac{11\pi}{6}\right) = -\frac{1}{2}$ and $\sin\left(\frac{\pi}{6}\right) = \frac{1}{2}$; $\cos\left(\frac{11\pi}{6}\right) = \frac{\sqrt{3}}{2}$ and $\cos\left(\frac{\pi}{6}\right) = \frac{\sqrt{3}}{2}$, etc. In other words, there is a close connection between the trigonometric values of an angle in standard position and its related reference angles. The only difference between the two angles involves occasional opposition of signs: Reference angles always produce positive values, but angles in standard position sometimes produce negative values. The quick way to deal with that issue is the mnemonic device "**A**ll **S**tudents **T**ake **C**alculus"—shorthand for: All trigonometric functions are positive in Quadrant I; Sine is positive in Quadrant II; Tangent is positive in Quadrant III; and Cosine is positive in Quadrant IV.

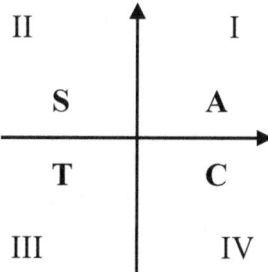

Now we can address the question of how to compute the value of the reference angles, e.g., $\sin\left(\frac{\pi}{6}\right) = \frac{1}{2}$. The two reference triangles come to our aid:

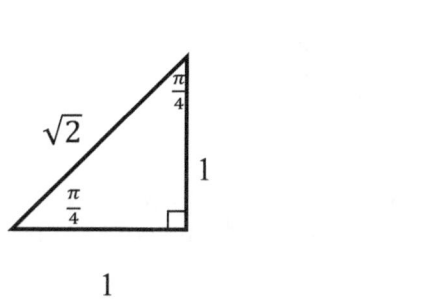 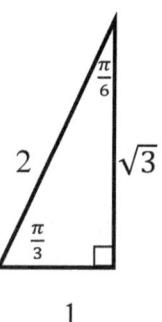

(Cf. pages 23-24 for the proper construction of these triangles.) We just read off from them the needed values. For example, for computing $\sin\left(\frac{\pi}{6}\right)$, look at the triangle that contains the angle with measure $\frac{\pi}{6}$ and apply SOHCAHTOA: $\sin\left(\frac{\pi}{6}\right) = \frac{\text{OPP}}{\text{HYP}} = \frac{1}{2}$.

To put all of this together, note the following procedure in the example where we compute the value of $\csc\left(\frac{5\pi}{3}\right)$.

1. $\csc\left(\frac{5\pi}{3}\right) = \frac{1}{\sin\left(\frac{5\pi}{3}\right)}$ and the reference angle for $\frac{5\pi}{3}$ is $\frac{\pi}{3}$.

2. Consulting the 30°-60°-90° triangle, we find that $\sin\left(\frac{\pi}{3}\right) = \frac{\sqrt{3}}{2}$. Hence, $\csc\left(\frac{\pi}{3}\right) = \frac{2}{\sqrt{3}}$.

3. But $\frac{5\pi}{3}$ is in Quadrant IV, and sine and cosecant are both negative there (ASTC).

4. Therefore, $\csc\left(\frac{5\pi}{3}\right) = -\frac{2}{\sqrt{3}}$.

This procedure obviates the need to consult the unit circle except for the special four points $(1,0)$, $(0,1)$, $(-1,0)$, $(0,-1)$, which correspond to the angles in standard position that have a measurement equal to a multiple of 90°.

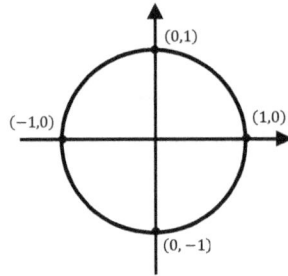

However, for any x-value on the unit circle, $x = \cos(\theta)$ and for any y-value $y = \sin(\theta)$, where θ is an angle in standard position. Hence, the values for cosine and sine corresponding to the special four points can be computed easily.

$$(1,0) \quad \Rightarrow \quad \theta = 0 \quad \Rightarrow \quad \cos(0) = 1 \quad \sin(0) = 0$$

$$(0,1) \quad \Rightarrow \quad \theta = \frac{\pi}{2} \quad \Rightarrow \quad \cos\left(\frac{\pi}{2}\right) = 0 \quad \sin\left(\frac{\pi}{2}\right) = 1$$

$$(-1,0) \quad \Rightarrow \quad \theta = \pi \quad \Rightarrow \quad \cos(\pi) = -1 \quad \sin(\pi) = 0$$

$$(0,-1) \quad \Rightarrow \quad \theta = \frac{3\pi}{2} \quad \Rightarrow \quad \cos\left(\frac{3\pi}{2}\right) = 0 \quad \sin\left(\frac{3\pi}{2}\right) = -1$$

"Only he who never plays, never loses"

www.ingramcontent.com/pod-product-compliance
Lightning Source LLC
Chambersburg PA
CBHW080614190526
45169CB00007B/3004